Global Weather

by BARBARA M. LINDE

Table of Contents

Chapter 1
What Is Weather?................2

Chapter 2
Wind and the Water Cycle......6

Chapter 3
Getting Colder,
Getting Warmer.................12

Chapter 4
Storms on the Planet...........16

Chapter 5
Snowstorms and Blizzards......24

Chapter 6
Dry as a Bone...................26

Chapter 7
Meet a Meteorologist...........28

Glossary..........................31

Index.............................32

Chapter 1
What Is Weather?

No matter where you are in the world, weather surrounds you. Weather changes from season to season. It can change from place to place, too. Students in Arizona may go to school on a hot, dry day. On that same day, children in London may go to school during a rainstorm. Workers in northern Canada may be shoveling the streets after a snowstorm. At the same time, farmers in Australia may be planting crops in sunny fields.

Weather is the condition of Earth's atmosphere. The atmosphere has layers of air. Think of it as a large blanket around Earth. The layers of the atmosphere extend hundreds of kilometers above Earth's surface. Global weather takes place in the lowest layer. This layer is called the **troposphere**. It begins at Earth's surface. It stretches upward from 8–18 kilometers (about 5–11 miles). The air in the troposphere stays in motion. This movement of air causes weather.

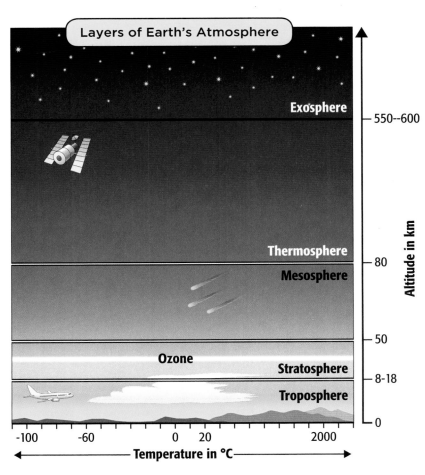

The Sun, Earth, air, and water work together to make global weather. Energy from the Sun is called **solar energy**. Solar energy travels to Earth as radiation, or heat rays. These rays warm Earth's atmosphere. Some areas on Earth get more energy than others. That is because Earth is tilted and in motion all the time. The area around the equator gets the most solar energy. The North and South Poles get the least energy.

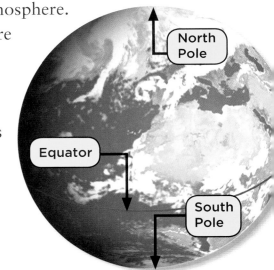

Solar energy travels from the Sun to Earth.

Air moves around because of differences in heat. Warm air moves from the equator toward the poles. Cool air moves from the poles toward the equator.

The air in the atmosphere has weight. This is called **air pressure**. When air pressure changes, weather usually changes with it. If air pressure falls, the effect is wet or windy weather. When air pressure rises, the sky becomes sunny and clear.

When air pressure falls, storms may form.

Chapter 2
Wind and the Water Cycle

The total amount of water on Earth is always the same. Most water is stored as a liquid in oceans or other bodies of water. Water is also stored underground. Some water is stored as ice at the polar ice caps and in glaciers. The rest is stored in the atmosphere as a gas called water vapor.

Precipitation
(Rain, snow, sleet, hai

Groundwater
(Water soaks into the ground)

Runoff
(Water flows back in rivers, lakes, and ocea

Water travels in a kind of circle called the **water cycle**. As water moves across Earth, it changes form. Energy from the Sun shines down on the surface water and heats it. The heated water **evaporates**. It changes into water vapor—a gas. The water vapor rises into the atmosphere. There, it **condenses** by changing from a gas into water drops.

These drops of water form around tiny particles of dust, salt, or sand in the air. As the drops group together, they form clouds. When the clouds get full, some of the water inside them falls to the ground. If the air is warm, it rains. If the air is cold, the water becomes snow, sleet, or hail. Once the water reaches Earth's surface, the water cycle begins again!

The Water Cycle

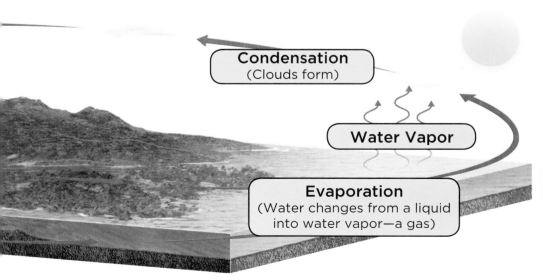

Condensation
(Clouds form)

Water Vapor

Evaporation
(Water changes from a liquid into water vapor—a gas)

Wind Patterns

Air moves from areas of high pressure to areas of low pressure. Winds travel all over Earth at different heights in the atmosphere. The winds balance Earth's temperature and air pressure. The movement is called **global wind patterns**.

There are three main groups of winds. Polar easterlies flow away from the poles. These winds are cold. Prevailing westerlies blow from the west between the equator and the poles. They carry warm air toward the poles and cool air toward the equator. Tropical easterlies, or trade winds, blow from the east through tropical regions close to the equator. Trade winds are usually mild and warm.

Some winds bring cold weather.

Winds move large amounts of air around the globe. If two of these air masses meet, they form a weather front. Fronts can produce wind, rain, or terrible storms. They also cause changes in temperature.

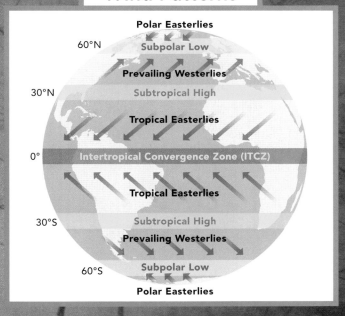

Ocean Currents

Oceans store energy as heat. Some of this heat rises and warms the atmosphere. This produces winds that blow across the surface of water. These winds also produce ocean currents. Ocean currents move water through the oceans. Like winds, ocean currents help move heat around Earth.

Every three to seven years, Earth's ocean currents change. The change begins in the cold waters of the eastern Pacific Ocean. It takes place near the coast of Peru, in South America. There, the ocean warms up in December. During some years, the ocean warms up much more than usual. This warmth produces the **El Niño** effect.

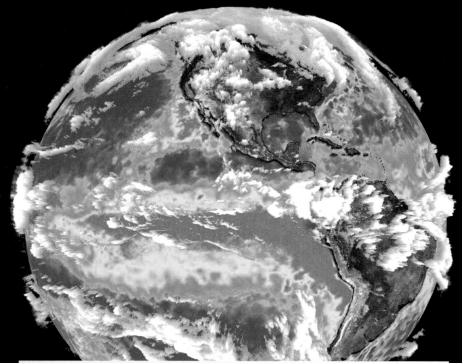

This is a computerized image of Earth. Earth's temperature is shown in colors from red (hot) to blue (cold). The clouds can be seen in three-dimensions.

El Niño caused terrible floods in Ecuador in 1998.

In 1998 El Niño caused floods in California.

El Niño affects weather all over Earth. The warmer water causes changes in global wind patterns. Areas that are usually dry get large amounts of rain and flooding. Areas that usually get a lot of rain suddenly get none. Cold areas warm up. Warm areas cool down.

Chapter 3
Getting Colder, Getting Warmer

About a million years ago, Earth's climate changed. Earth's temperature dropped. Temperatures continued to fall for thousands of years. Large glaciers formed. They covered large parts of Earth's surface. This was the Great Ice Age.

The Great Ice Age affected our planet. As the glaciers moved, they changed the surface of the land. They carved deep valleys. They forced rivers to change course. They even wore down mountains!

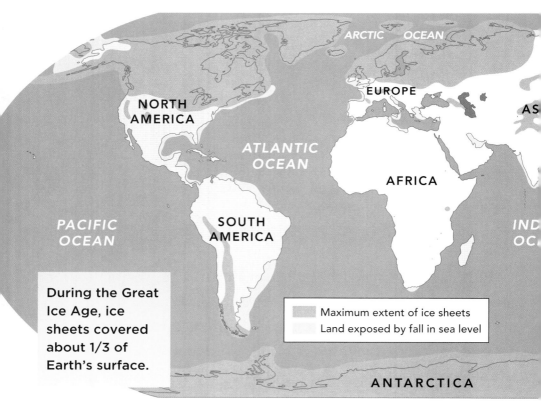

During the Great Ice Age, ice sheets covered about 1/3 of Earth's surface.

Maximum extent of ice sheets
Land exposed by fall in sea level

About 20,000 years ago, Earth slowly began warming. Ice sheets melted. The level of water in the oceans rose. Holes carved by glaciers filled with water. These became lakes. When glaciers melted, they left rocks, sand, and gravel behind.

Scientists believe the Great Ice Age finally ended about 6,000 years ago. Will Earth ever face another ice age? Scientists are not sure. But some think it could happen in the future.

Glaciers produced valleys such as this one during the Great Ice Age.

Global Warming

Earth's average temperature has risen about one degree during the past hundred years. This small rise in temperature is called **global warming**. Some global warming may be natural. Yet human actions may play a role, too.

Here is what scientists know. Gases, including carbon dioxide, keep heat in the atmosphere and warm Earth. This is usually a good thing. But many scientists now think there are too many gases in the atmosphere.

Air pollution may help cause global warming.

Cutting down large areas of forests may help cause global warming.

People use machines that produce air pollution. This causes a rise in carbon dioxide, a harmful gas. Large areas of forests are cut down to make room for buildings and farmland. The trees in these forests help control carbon dioxide levels. If there are fewer trees, then there will be more carbon dioxide in the air.

What might happen if Earth's temperature continues to rise? There may be changes in the amount of rainfall. Sea levels may rise. Plants, animals, and people could be in danger. As scientists learn more, they will gain a better understanding of the causes and effects of global warming.

Chapter 4

Storms on the Planet

You probably have heard loud booms of thunder. You probably have seen flashes of lightning in the sky. About 1,800 thunderstorms take place on Earth every day! Most last about 30 minutes. These thunderstorms cover an area of about 24 kilometers (15 miles).

A thunderstorm begins with a pocket of warm, moist air in the lower part of the atmosphere. When cold air pushes this warm air up into the atmosphere, it cools off. Then it condenses and forms clouds. The thunderclouds grow and get heavier. Finally, they release rain.

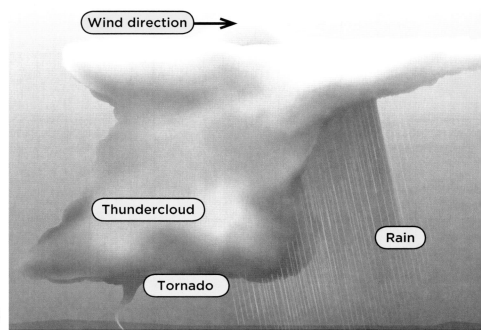

Wind direction →

Thundercloud

Rain

Tornado

16

The air inside a thundercloud moves quickly. This causes drops of water in the cloud to crash into each other. This produces electricity. The electricity makes sparks that form a flash of lightning. Sometimes the lightning jumps from the cloud and strikes the ground.

The air around a bolt of lightning heats up quickly. Then it expands. As it grows larger, it sends out sound waves. These sound waves are thunder!

This flash of lightning strikes above the Eiffel Tower in Paris, France.

Tornadoes

A tornado looks like a large funnel hanging down from a thundercloud. Its funnel shape is made by rising warm air. The air forms a column in the center of the thunderstorm. Winds make the column spin at speeds of up to 480 kilometers (298 miles) an hour!

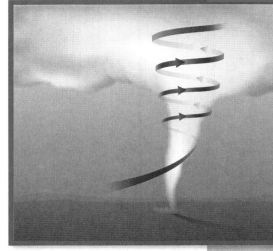

A tornado begins in a thunderstorm and moves down to the ground.

As a tornado spins, it moves along the ground, making a loud roaring noise. Like a giant vacuum cleaner, it can suck up anything in its path—trash cans, cars, and even houses! These items swirl in the funnel. Then they are dropped to the ground. Some items end up miles away. A tornado's path can be wild and hard to predict. Often it will destroy one house but not touch the one next to it!

Tornadoes are violent storms. Most of the world's tornadoes happen in the United States, east of the Rocky Mountains. Most take place during the spring and summer. These dangerous storms may not last long. Yet the damage from tornadoes is often great.

Tornadoes can cause terrible damage in just seconds.

Hurricanes

A hurricane has winds of at least 119 kilometers (74 miles) an hour. It can also have hard rain and thunderstorms. In the Pacific region, these storms are called typhoons or cyclones. Whatever their name, these storms are loud and scary!

Hurricanes form above warm tropical waters. They collect heat, energy, and moisture from the warm water. This energy produces winds that flow in a spiral pattern. Thunderstorms form and rise upward. Then they spiral around the center of the storm. The trade winds push the storm over the ocean, where the hurricane gains in strength and size.

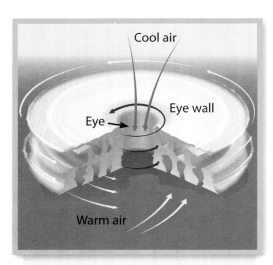

This diagram shows how a hurricane forms. The eye in the center is calm. The strongest winds are in the eyewall that surrounds the eye.

If a hurricane strikes land, it can cause great damage. Strong winds can uproot trees and destroy buildings. These winds can also cause big ocean waves. This is called a storm surge. Rain can cause flooding. In time, hurricanes get weak and die out. Yet their effects may last for years.

Hurricane Katrina was a large, powerful storm. It hit the Gulf Coast of the United States in August 2005 and caused terrible flooding.

Monsoons

The word **monsoon** is an Arabic word meaning "the season of winds." In areas that have monsoons, winds blow in from one direction for part of the year. They blow in another direction for the other part of the year. The difference in temperature between land and sea causes the wind change.

In Southeast Asia, monsoon winds come from the northeast during the winter. These winds are hot and dry. The Himalaya Mountains block out any cold air and moisture. Summer monsoon winds come from the southwest. These winds flow over the Indian Ocean and gather large amounts of moisture. From about May until September, this moisture falls as rain. These summer rains are heavy. They can cause flooding.

Mexico also has monsoon winds. They usually happen from June through October. Most of Mexico's rain falls during this time. Some of this wind and rain may make its way to Arizona and New Mexico.

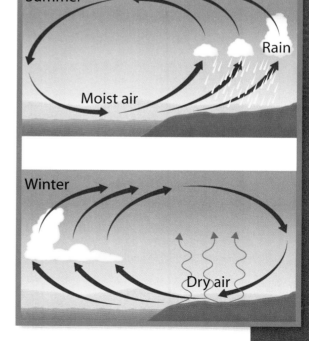

Southeast Asia Monsoons

Flooding accompanies the wet monsoon winds.

Chapter 5

Snowstorms and Blizzards

Sometimes snowstorms are fun. You can ski or sled on the powder. You can catch the floating white flakes on your tongue. How do these storms form?

First, cold northern air flows southward. When moisture falls from the clouds into the cold air, snowflakes form and fall to Earth. If the ground temperature is below freezing—0° Celsius (32° Fahrenheit)—then the snow sticks to the ground. The snow melts when the temperature rises above freezing.

A blizzard is a severe snowstorm. A snowstorm becomes a blizzard if it lasts for more than three straight hours and has temperatures below freezing. It also must have gusts of wind blowing at least 56 kilometers (35 miles) an hour. In a blizzard, large amounts of snow swirl around and fall to the ground.

A blizzard forms on the northwest side of a storm system. Low pressure in the storm and high pressure in the west cause strong winds. In time, these winds merge with the falling snow.

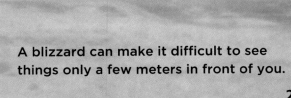

A blizzard can make it difficult to see things only a few meters in front of you.

Chapter 6

Dry as a Bone

Many severe weather systems drop large amounts of rain or snow. Yet a **drought** is different. During a drought, there is not enough water. The amount of rain or snowfall is less than average for the area. Some droughts last for a short time. Others can continue for years. During a drought, there are often high temperatures, high winds, and low humidity.

A harsh drought caused the "Dust Bowl" in the 1930s.

A drought can affect Earth in many ways. During a drought, bodies of water fall below normal levels. Soil dries up. Crops and other plants die from not getting enough water. When the plants die, there is nothing to protect the soil beneath them. The loose soil may blow around and cause dust storms. Dead or dying plants also burn easily. This can lead to fires. Drying forests may begin to look like deserts. During a drought, it is important to save water.

A drought destroyed this corn crop.

Chapter 7
Meet a Meteorologist

Weather forecasters study Earth's atmosphere, climate, and weather. These scientists are called **meteorologists**.

Dr. Laura Hinkelman is a meteorologist. Clouds are one reason she loves her job.

Dr. Hinkelman studies clouds and Earth's energy budget. The energy budget

Dr. Hinkelman works for the National Institute of Aerospace.

is the total amount of solar energy coming into Earth's atmosphere. It is also the amount of heat energy leaving the atmosphere. These amounts have to be in balance. If they are not, the results can affect all of Earth.

"Increased heating or cooling could cause changes in the atmosphere," Dr. Hinkelman says. "We still don't understand a lot about clouds. Will there be more types of clouds if the atmosphere heats up? Will there be fewer? A change in the clouds could alter the amount of rainfall on Earth. It could change the amount of solar energy that reaches Earth."

Satellites measure the energy high in the atmosphere. People on the ground make surface measurements. Dr. Hinkelman looks for patterns of change. "There are many natural and human factors to consider," she says. Meteorologists "want to know what is happening."

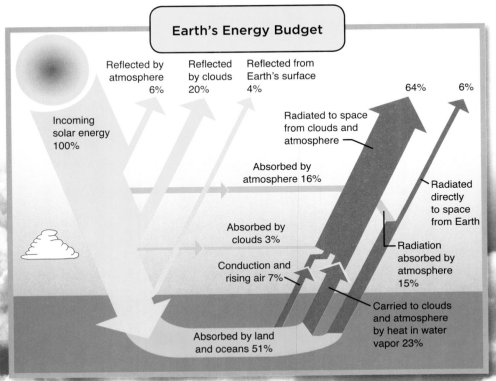

This diagram of Earth's energy budget shows how thermal and solar energy travel through the atmosphere.

It is exciting to learn what the weather is like in different parts of Earth. If you travel a lot, you may see many kinds of mild and severe weather. The more you know about weather, the more you will appreciate the wonders of weather around the globe.

30

Glossary

air pressure (AYR PRESH-uhr) the force put on an area by the weight of the air above it *(page 5)*

condense (kuhn-DENS) to change from a gas to a liquid *(page 7)*

drought (DROWT) a long period of time with little rain or snow *(page 26)*

El Niño (EL NEEN-yoh) an unusual warming in the surface temperature in part of the Pacific Ocean, affecting the world's weather *(page 10)*

evaporate (i-VAP-uh-rayt) to change from a liquid to a gas *(page 7)*

global warming (GLOH-buhl WORM-ing) a rise in Earth's average temperature *(page 14)*

global wind patterns (GLOH-buhl WIND PAT-urnz) movement of air around Earth *(page 8)*

meteorologist (mee-tee-uh-ROL-uh-jist) a scientist who studies Earth's weather *(page 28)*

monsoon (mahn-SOON) winds that reverse directions with the seasons *(page 22)*

solar energy (SOH-luhr EN-uhr-jee) power from sunlight *(page 4)*

troposphere (TROP-uh-sfeer) the layer of Earth's atmosphere where weather happens *(page 3)*

water cycle (WAW-tuhr SI-kuhl) movement of Earth's water from a liquid to a gas and back *(page 7)*

Index

air pressure, 5, 8

atmosphere, 3-6, 8, 10, 14, 16, 28-29

blizzard, 24-25

condense, 7, 16

cyclone, 20

drought, 26-27

El Niño, 10-11

energy budget, 28-29

evaporate, 7, 16

glacier, 6, 12-13

global warming, 14-15

global wind patterns, 8-9, 11

Great Ice Age, 12-13

hail, 6-7

Hinkelman, Dr. Laura, 28-29

hurricane, 20-21

lightning, 16-17

meteorologist, 28-29

monsoon, 22-23

ocean currents, 10

rain, 2, 6-7, 11, 16, 20, 22, 26, 29

sleet, 6-7

snow, 6-7, 24-26

snowstorm, 2, 24-25

solar energy, 4, 28-29

thunderstorm, 16, 18, 20

tornado, 18-19

troposphere, 3

typhoon, 20

water cycle, 6-7

water vapor, 6-7

weather, 2-5, 26, 28, 30